This is a printed compilation for people that enjoy using and working with printed manuals. The information in this compilation is available for free in PDF format directly from Raspberry Pi. This manual is printed in accordance with their CC BY-ND license. This is a third party printing of their documentation by DienstNet LLC. As an extension, all parts of this compilation not covered by the Raspberry Pi license are also provided under the same CC-BY-ND copyright by DienstNet LLC 2023.

Compilation Contents
Raspberry Pi Pico W Pinout
https://datasheets.raspberrypi.com/pico/PicoW-R3-A4-Pinout.pdf

Raspberry Pi Pico W Product Brief
https://datasheets.raspberrypi.com/pico/pico-w-product-brief.pdf

Raspberry Pi Pico W Datasheet
https://datasheets.raspberrypi.com/pico/pico-w-datasheet.pdf

ISBN 978-1-365-06659-7

This page was intentionally left blank.

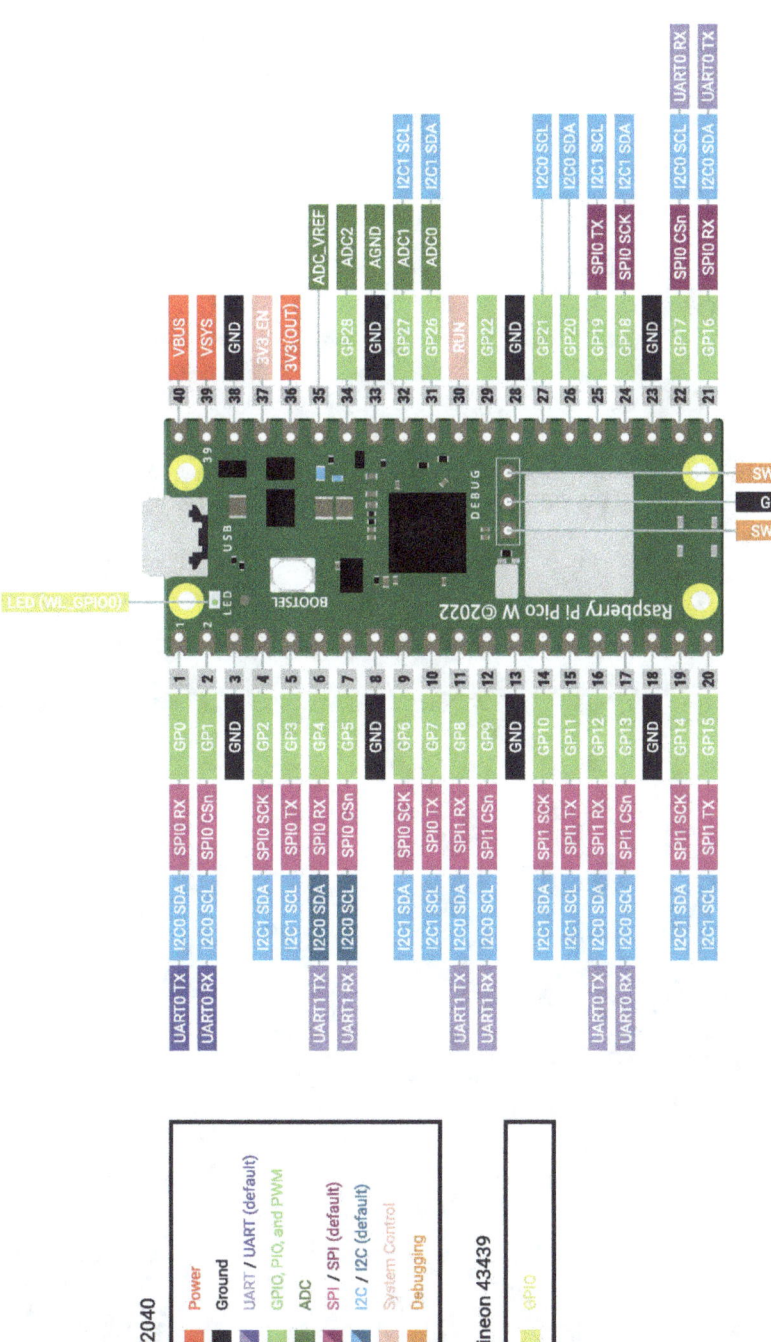

This page was intentionally left blank.

Raspberry Pi Pico W

Published June 2022

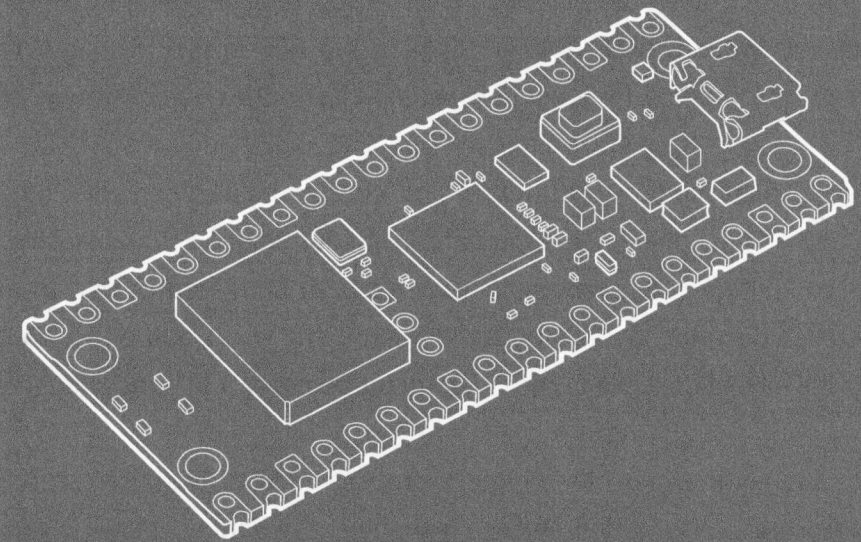

Raspberry Pi Ltd

Overview

Raspberry Pi Pico W brings wireless connectivity to the best-selling Raspberry Pi Pico product line. Built around our RP2040 silicon platform, Pico products bring our signature values of high performance, low cost, and ease of use to the microcontroller space.

With a large on-chip memory, symmetric dual-core processor complex, deterministic bus fabric, and rich peripheral set augmented with our unique Programmable I/O (PIO) subsystem, RP2040 provides professional users with unrivalled power and flexibility. Offering detailed documentation, a polished MicroPython port, and a UF2 bootloader in ROM, it has the lowest possible barrier to entry for beginner and hobbyist users.

RP2040 is manufactured on a modern 40nm process node, delivering high performance, low dynamic power consumption, and low leakage, with a variety of low-power modes to support extended-duration operation on battery power.

Raspberry Pi Pico W offers 2.4GHz 802.11 b/g/n wireless LAN support, with an on-board antenna, and modular compliance certification. It is able to operate in both station and access-point modes. Full access to network functionality is available to both C and MicroPython developers.

Raspberry Pi Pico W pairs RP2040 with 2MB of flash memory, and a power supply chip supporting input voltages from 1.8–5.5V. It provides 26 GPIO pins, three of which can function as analogue inputs, on 0.1"-pitch through-hole pads with castellated edges. Raspberry Pi Pico W is available as an individual unit, or in 480-unit reels for automated assembly.

Specification

Form factor:	21 mm × 51 mm
CPU:	Dual-core Arm Cortex-M0+ @ 133MHz
Memory:	264KB on-chip SRAM; 2MB on-board QSPI flash
Interfacing:	26 GPIO pins, including 3 analogue inputs
Peripherals:	• 2 × UART • 2 × SPI controllers • 2 × I2C controllers • 16 × PWM channels • 1 × USB 1.1 controller and PHY, with host and device support • 8 × PIO state machines
Connectivity:	2.4GHz IEEE 802.11b/g/n wireless LAN, on-board antenna
Input power:	1.8–5.5V DC
Operating temperature:	-20°C to +70°C
Production lifetime:	Raspberry Pi Pico W will remain in production until at least January 2028
Compliance:	For a full list of local and regional product approvals, please visit pip.raspberrypi.com

Physical specification

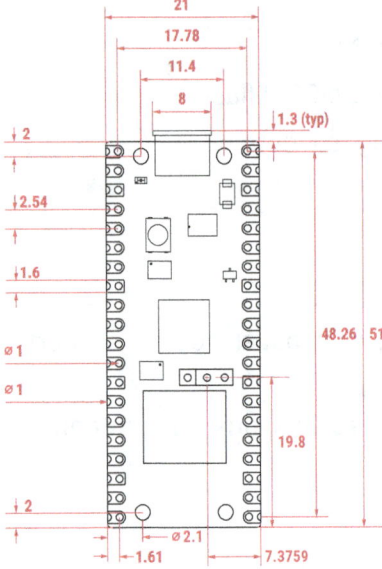

Note: all dimensions in mm

WARNINGS

- Any external power supply used with Raspberry Pi Pico W shall comply with relevant regulations and standards applicable in the country of intended use.
- This product should be operated in a well-ventilated environment, and if used inside a case, the case should not be covered.
- Whilst in use, this product should be placed on a stable, flat, non-conductive surface, and should not be contacted by conductive items.
- The connection of incompatible devices to Raspberry Pi Pico W may affect compliance, result in damage to the unit, and invalidate the warranty.
- All accessories used with this product should comply with relevant standards for the country of use and be marked accordingly to ensure that safety and performance requirements are met.
- The cables and connectors of all peripherals used with this product must have adequate insulation so that relevant safety requirements are met.

SAFETY INSTRUCTIONS

To avoid malfunction or damage to this product, please observe the following:

- Do not expose to water or moisture, or place on a conductive surface whilst in operation.
- Do not expose to heat from any source; Raspberry Pi Pico W is designed for reliable operation at normal ambient temperatures.
- Take care whilst handling to avoid mechanical or electrical damage to the printed circuit board and connectors.
- Whilst it is powered, avoid handling the printed circuit board, or only handle it by the corners to minimise the risk of electrostatic discharge damage.

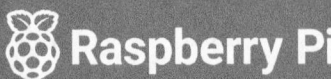

Raspberry Pi is a trademark of Raspberry Pi Ltd

Raspberry Pi Pico W Datasheet
An RP2040-based microcontroller board with wireless

Colophon

Copyright © 2022 Raspberry Pi Ltd

The documentation of the RP2040 microcontroller is licensed under a Creative Commons Attribution-NoDerivatives 4.0 International (CC BY-ND).

build-date: 2022-11-30
build-version: 3a2defe-clean

> **About the SDK**
>
> Throughout the text "the SDK" refers to our Raspberry Pi Pico SDK. More details about the SDK can be found in the Raspberry Pi Pico C/C++ SDK book. Source code included in the documentation is Copyright © 2020-2022 Raspberry Pi Ltd (formerly Raspberry Pi (Trading) Ltd.) and licensed under the 3-Clause BSD license.

Legal disclaimer notice

TECHNICAL AND RELIABILITY DATA FOR RASPBERRY PI PRODUCTS (INCLUDING DATASHEETS) AS MODIFIED FROM TIME TO TIME ("RESOURCES") ARE PROVIDED BY RASPBERRY PI LTD ("RPL") "AS IS" AND ANY EXPRESS OR IMPLIED WARRANTIES, INCLUDING, BUT NOT LIMITED TO, THE IMPLIED WARRANTIES OF MERCHANTABILITY AND FITNESS FOR A PARTICULAR PURPOSE ARE DISCLAIMED. TO THE MAXIMUM EXTENT PERMITTED BY APPLICABLE LAW IN NO EVENT SHALL RPL BE LIABLE FOR ANY DIRECT, INDIRECT, INCIDENTAL, SPECIAL, EXEMPLARY, OR CONSEQUENTIAL DAMAGES (INCLUDING, BUT NOT LIMITED TO, PROCUREMENT OF SUBSTITUTE GOODS OR SERVICES; LOSS OF USE, DATA, OR PROFITS; OR BUSINESS INTERRUPTION) HOWEVER CAUSED AND ON ANY THEORY OF LIABILITY, WHETHER IN CONTRACT, STRICT LIABILITY, OR TORT (INCLUDING NEGLIGENCE OR OTHERWISE) ARISING IN ANY WAY OUT OF THE USE OF THE RESOURCES, EVEN IF ADVISED OF THE POSSIBILITY OF SUCH DAMAGE.

RPL reserves the right to make any enhancements, improvements, corrections or any other modifications to the RESOURCES or any products described in them at any time and without further notice.

The RESOURCES are intended for skilled users with suitable levels of design knowledge. Users are solely responsible for their selection and use of the RESOURCES and any application of the products described in them. User agrees to indemnify and hold RPL harmless against all liabilities, costs, damages or other losses arising out of their use of the RESOURCES.

RPL grants users permission to use the RESOURCES solely in conjunction with the Raspberry Pi products. All other use of the RESOURCES is prohibited. No licence is granted to any other RPL or other third party intellectual property right.

HIGH RISK ACTIVITIES. Raspberry Pi products are not designed, manufactured or intended for use in hazardous environments requiring fail safe performance, such as in the operation of nuclear facilities, aircraft navigation or communication systems, air traffic control, weapons systems or safety-critical applications (including life support systems and other medical devices), in which the failure of the products could lead directly to death, personal injury or severe physical or environmental damage ("High Risk Activities"). RPL specifically disclaims any express or implied warranty of fitness for High Risk Activities and accepts no liability for use or inclusions of Raspberry Pi products in High Risk Activities.

Raspberry Pi products are provided subject to RPL's Standard Terms. RPL's provision of the RESOURCES does not expand or otherwise modify RPL's Standard Terms including but not limited to the disclaimers and warranties expressed in them.

Table of contents

Colophon 1
    Legal disclaimer notice 1
1. About Raspberry Pi Pico W 3
    1.1. Raspberry Pi Pico W design files 5
2. Mechanical specification 7
    2.1. Pico W pinout 7
    2.2. Surface-mount footprint 9
    2.3. Recommended operating conditions 10
3. Applications information 11
    3.1. Programming the flash 11
    3.2. General purpose I/O 11
    3.3. Using the ADC 11
    3.4. Powerchain 12
    3.5. Powering Raspberry Pi Pico W 13
    3.6. Using a battery charger 14
    3.7. USB 15
    3.8. Wireless interface 15
    3.9. Debugging 16
Appendix A: Availability 17
    Support 17
    Ordering code 17
Appendix B: Pico W schematic 18
Appendix C: Pico W component locations 20
Appendix D: Documentation release history 21

Chapter 1. About Raspberry Pi Pico W

Raspberry Pi Pico W is a microcontroller board based on the Raspberry Pi RP2040 microcontroller chip.

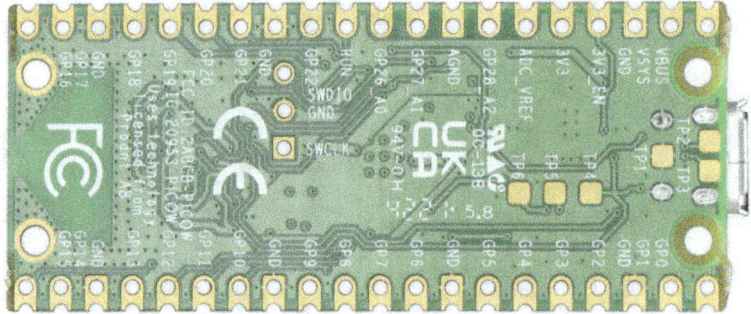

Figure 1. The Raspberry Pi Pico W Rev3 board.

Raspberry Pi Pico W has been designed to be a low cost yet flexible development platform for RP2040, with a 2.4GHz wireless interface and the following key features:

- RP2040 microcontroller with 2MB of flash memory
- On-board single-band 2.4GHz wireless interfaces (802.11n)
- Micro USB B port for power and data (and for reprogramming the flash)
- 40-pin 21mm×51mm 'DIP' style 1mm thick PCB with 0.1" through-hole pins also with edge castellations
    - Exposes 26 multi-function 3.3V general purpose I/O (GPIO)
    - 23 GPIO are digital-only, with three also being ADC capable
    - Can be surface-mounted as a module
- 3-pin Arm serial wire debug (SWD) port
- Simple yet highly flexible power supply architecture
    - Various options for easily powering the unit from micro USB, external supplies or batteries
- High quality, low cost, high availability
- Comprehensive SDK, software examples and documentation

For full details of the RP2040 microcontroller please see the RP2040 Datasheet book. Key features include:

- Dual-core cortex M0+ at up to 133MHz
    - On-chip PLL allows variable core frequency
- 264kB multi-bank high performance SRAM

- External Quad-SPI flash with eXecute In Place (XIP) and 16kB on-chip cache
- High performance full-crossbar bus fabric
- On-board USB1.1 (device or host)
- 30 multi-function general purpose I/O (four can be used for ADC)
    - 1.8-3.3V I/O voltage
- 12-bit 500ksps analogue to digital converter (ADC)
- Various digital peripherals
    - 2 × UART, 2 × I2C, 2 × SPI, 16 × PWM channels
    - 1 × timer with 4 alarms, 1 × real time clock
- 2 × programmable I/O (PIO) blocks, 8 state machines in total
    - Flexible, user-programmable high-speed I/O
    - Can emulate interfaces such as SD card and VGA

> **NOTE**
>
> Raspberry Pi Pico W I/O voltage is fixed at 3.3V

Raspberry Pi Pico W provides a minimal yet flexible external circuitry to support the RP2040 chip: flash memory (Winbond W25Q16JV), a crystal, power supplies and decoupling, and USB connector. The majority of the RP2040 microcontroller pins are brought to the user I/O pins on the left and right edge of the board. Four RP2040 I/O are used for internal functions: driving an LED, on-board switch mode power supply (SMPS) power control, and sensing the system voltages.

Pico W has an on-board 2.4GHz wireless interface using an Infineon CYW43439. The antenna is an onboard antenna licensed from ABRACON (formerly ProAnt). The wireless interface is connected via SPI to the RP2040.

Pico W has been designed to use either soldered 0.1-inch pin-headers (it is one 0.1-inch pitch wider than a standard 40-pin DIP package), or to be positioned as a surface-mountable 'module', as the user I/O pins are also castellated. There are SMT pads underneath the USB connector and BOOTSEL button, which allow these signals to be accessed if used as a reflow-soldered SMT module.

Figure 2. The pinout of the Pico W Rev3 board

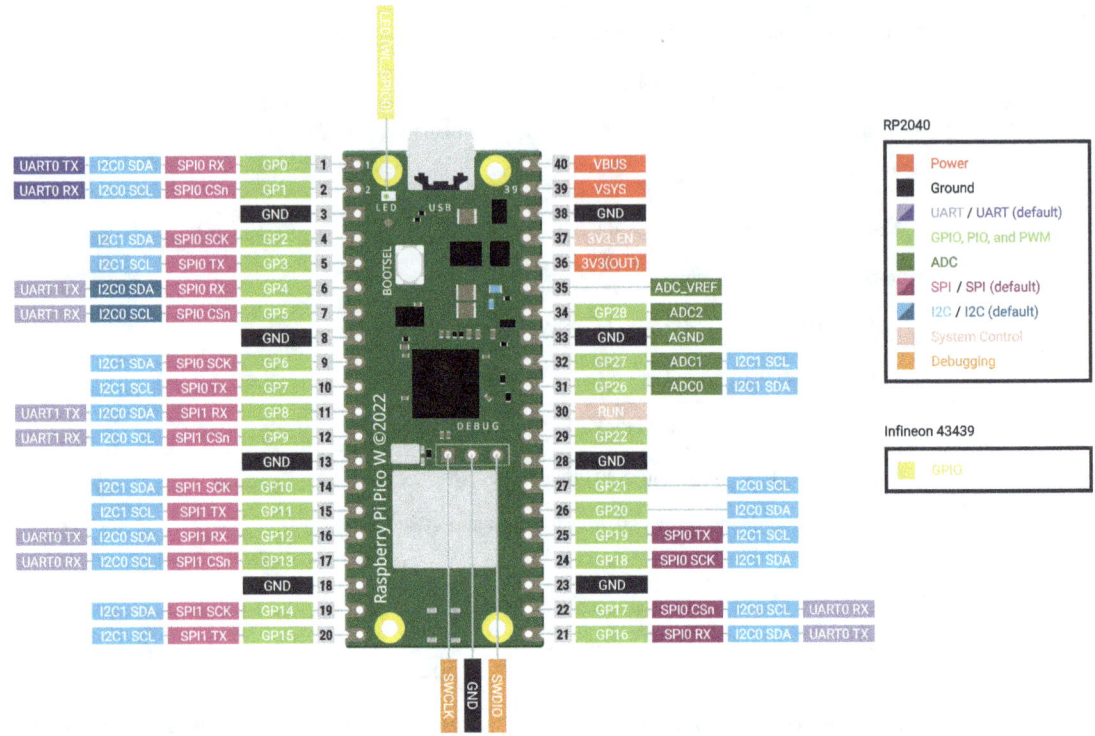

Raspberry Pi Pico W uses an on-board buck-boost SMPS which is able to generate the required 3.3V (to power RP2040 and external circuitry) from a wide range of input voltages (~1.8 to 5.5V). This allows significant flexibility in powering the unit from various sources, such as a single lithium-ion cell, or three AA cells in series. Battery chargers can also be very easily integrated with the Pico W powerchain.

Reprogramming the Pico W flash can be done using USB (simply drag and drop a file onto the Pico W, which appears as a mass storage device), or the standard serial wire debug (SWD) port can reset the system and load and run code without any button presses. The SWD port can also be used to interactively debug code running on the RP2040.

Getting started with Pico W

The Getting started with Raspberry Pi Pico book walks through loading programs onto the board, and shows how to install the C/C++ SDK and build the example C programs. See the Raspberry Pi Pico Python SDK book to get started with MicroPython, which is the fastest way to get code running on Pico W.

1.1. Raspberry Pi Pico W design files

The source design files, including the schematic and PCB layout, are made available openly except for the antenna. The Niche™ antenna is an Abracon/Proant patented antenna technology. Please contact niche@abracon.com for information on licensing.

Schematic	The schematic is reproduced in Appendix B. The schematic is also distributed alongside the layout files.
Layout	The CAD files, including PCB layout, can be found here. Note that Pico W was designed in Cadence Allegro PCB Editor, and opening in other PCB CAD packages will require an import script or plugin.

STEP 3D   A STEP 3D model of Raspberry Pi Pico W, for 3D visualisation and fit-check of designs which include Pico W as a module, can be found here.

Fritzing   A Fritzing part for use in e.g. breadboard layouts can be found here.

Permission to use, copy, modify, and/or distribute this design for any purpose with or without fee is hereby granted.

THE DESIGN IS PROVIDED "AS IS" AND THE AUTHOR DISCLAIMS ALL WARRANTIES WITH REGARD TO THIS DESIGN INCLUDING ALL IMPLIED WARRANTIES OF MERCHANTABILITY AND FITNESS. IN NO EVENT SHALL THE AUTHOR BE LIABLE FOR ANY SPECIAL, DIRECT, INDIRECT, OR CONSEQUENTIAL DAMAGES OR ANY DAMAGES WHATSOEVER RESULTING FROM LOSS OF USE, DATA OR PROFITS, WHETHER IN AN ACTION OF CONTRACT, NEGLIGENCE OR OTHER TORTIOUS ACTION, ARISING OUT OF OR IN CONNECTION WITH THE USE OR PERFORMANCE OF THIS DESIGN.

Chapter 2. Mechanical specification

The Pico W is a single sided 51mm × 21mm × 1mm PCB with a micro USB port overhanging the top edge, and dual castellated/through-hole pins around the two long edges. The onboard wireless antenna is located on the bottom edge. To avoid detuning the antenna, no material should intrude into this space. Pico W is designed to be usable as a surface-mount module as well as presenting a dual inline package (DIP) format, with the 40 main user pins on a 2.54mm (0.1") pitch grid with 1mm holes, compatible with veroboard and breadboard. Pico W also has four 2.1mm (± 0.05mm) drilled mounting holes to provide for mechanical fixing(see Figure 3).

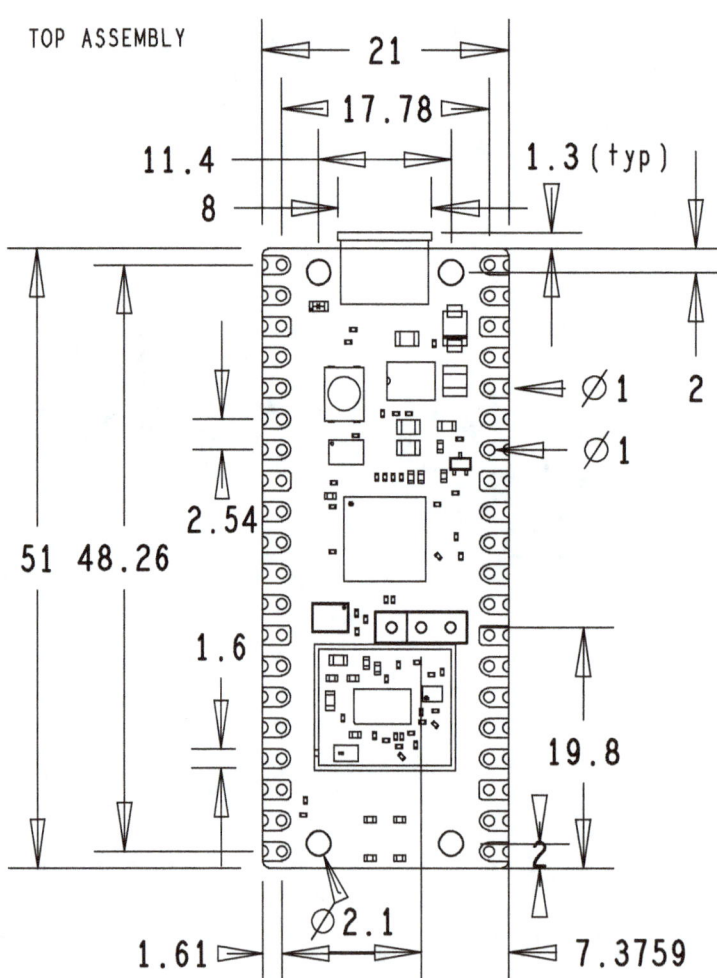

Figure 3. The dimensions of the Pico W

2.1. Pico W pinout

The Pico W pinout has been designed to directly bring out as much of the RP2040 GPIO and internal circuitry function as possible, while also providing a suitable number of ground pins to reduce electro-magnetic interference (EMI) and signal crosstalk. RP2040 is built on a modern 40nm silicon process, so its digital I/O edge rates are very fast.

Figure 4. The pin numbering of the Pico W

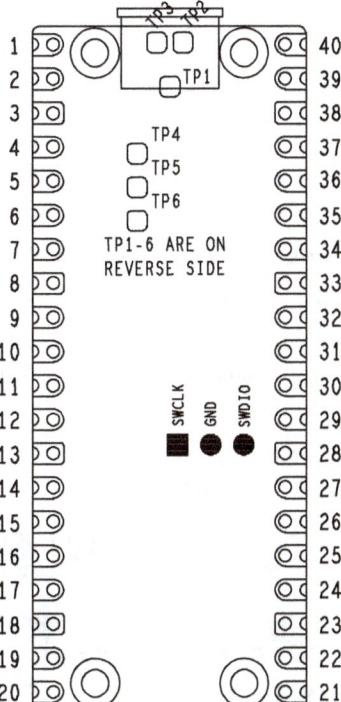

> **ⓘ NOTE**
>
> The physical pin numbering is shown in Figure 4. For pin allocation see Figure 2, or the full Pico W schematics in Appendix B.

A few RP2040 GPIO pins are used for internal board functions:

    GPIO29   OP/IP wireless SPI CLK/ADC mode (ADC3) to measure VSYS/3

    GPIO25   OP wireless SPI CS - when high also enables GPIO29 ADC pin to read VSYS

    GPIO24   OP/IP wireless SPI data/IRQ

    GPIO23   OP wireless power on signal

    WL_GPIO2   IP VBUS sense - high if VBUS is present, else low

    WL_GPIO1   OP controls the on-board SMPS power save pin (Section 3.4)

    WL_GPIO0   OP connected to user LED

Apart from GPIO and ground pins, there are seven other pins on the main 40-pin interface:

    PIN40   VBUS

    PIN39   VSYS

    PIN37   3V3_EN

    PIN36   3V3

PIN35 ADC_VREF

PIN33 AGND

PIN30 RUN

VBUS is the micro-USB input voltage, connected to micro-USB port pin 1. This is nominally 5V (or 0V if the USB is not connected or not powered).

VSYS is the main system input voltage, which can vary in the allowed range 1.8V to 5.5V, and is used by the on-board SMPS to generate the 3.3V for the RP2040 and its GPIO.

3V3_EN connects to the on-board SMPS enable pin, and is pulled high (to VSYS) via a 100kΩ resistor. To disable the 3.3V (which also de-powers the RP2040), short this pin low.

3V3 is the main 3.3V supply to RP2040 and its I/O, generated by the on-board SMPS. This pin can be used to power external circuitry (maximum output current will depend on RP2040 load and VSYS voltage; it is recommended to keep the load on this pin under 300mA).

ADC_VREF is the ADC power supply (and reference) voltage, and is generated on Pico W by filtering the 3.3V supply. This pin can be used with an external reference if better ADC performance is required.

AGND is the ground reference for GPIO26-29. There is a separate analogue ground plane running under these signals and terminating at this pin. If the ADC is not used or ADC performance is not critical, this pin can be connected to digital ground.

RUN is the RP2040 enable pin, and has an internal (on-chip) pull-up resistor to 3.3V of about ~50kΩ. To reset RP2040, short this pin low.

Finally, there are also six test points (TP1-TP6), which can be accessed if required, for example if using as a surface-mount module. These are:

TP1 Ground (close-coupled ground for differential USB signals)

TP2 USB DM

TP3 USB DP

TP4 WL_GPIO1/SMPS PS pin (do not use)

TP5 WL_GPIO0/LED (not recommended to be used)

TP6 BOOTSEL

TP1, TP2 and TP3 can be used to access USB signals instead of using the micro-USB port. TP6 can be used to drive the system into mass-storage USB programming mode (by shorting it low at power-up). Note that TP4 is not intended to be used externally, and TP5 is not really recommended to be used as it will only swing from 0V to the LED forward voltage (and hence can only really be used as an output with special care).

2.2. Surface-mount footprint

The following footprint (Figure 5) is recommended for systems which will be reflow-soldering Pico W units as modules.

Figure 5. The SMT footprint of the Pico W Rev3 board.

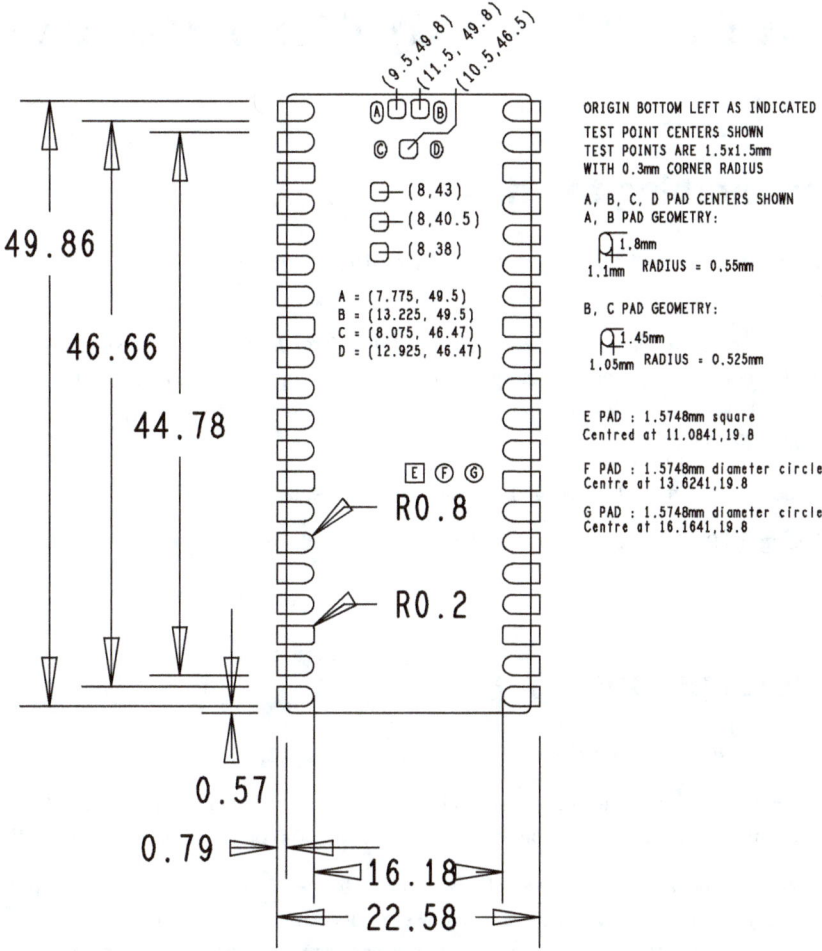

The footprint shows the test point locations and pad sizes as well as the 4 USB connector shell ground pads (A,B,C,D). The USB connector on Pico W is a through-hole part, which provides it with mechanical strength. The USB socket pins do not protrude all the way through the board, however solder does pool at these pads during manufacture and can stop the module sitting completely flat. Hence we provide pads on the SMT module footprint to allow this solder to reflow in a controlled manner when Pico W goes through reflow again.

For test points that are not used, it is acceptable to void any copper under these (with suitable clearance) on the carrier board.

2.3. Recommended operating conditions

Operating conditions for the Pico W are largely a function of the operating conditions specified by its components.

Operating Temp Max	70°C (including self-heating)
Operating Temp Min	-20°C
VBUS	5V ± 10%.
VSYS Min	1.8V
VSYS Max	5.5V

Note that VBUS and VSYS current will depend on use-case, some examples are given in the next section.

Recommended maximum ambient temperature of operation is 70°C.

Chapter 3. Applications information

3.1. Programming the flash

The on-board 2MB QSPI flash can be (re)programmed either using the serial wire debug port or by the special USB mass storage device mode.

The simplest way to reprogram the Pico W's flash is to use the USB mode. To do this, power-down the board, then hold the BOOTSEL button down during board power-up (e.g. hold BOOTSEL down while connecting the USB). The Pico W will then appear as a USB mass storage device. Dragging a special '.uf2' file onto the disk will write this file to the flash and restart the Pico W.

The USB boot code is stored in ROM on RP2040, so can not be accidentally overwritten.

To get started using the SWD port see the Debugging with SWD section in the Getting started with Raspberry Pi Pico book.

3.2. General purpose I/O

The Pico W's GPIO is powered from the on-board 3.3V rail, and is fixed at 3.3V.

Pico W exposes 26 of the 30 possible RP2040 GPIO pins by routing them straight out to Pico W header pins. GPIO0 to GPIO22 are digital only, and GPIO 26-28 can be used either as digital GPIO or as ADC inputs (software selectable).

GPIO 26-29 are ADC-capable and have an internal reverse diode to the VDDIO (3.3V) rail, so the input voltage must not exceed VDDIO plus about 300mV. If the RP2040 is unpowered, applying a voltage to these GPIO pins will 'leak' through the diode into the VDDIO rail. GPIO pins 0-25 (and the debug pins) do not have this restriction and therefore voltage can safely be applied to these pins when RP2040 is unpowered up to 3.3V.

3.3. Using the ADC

The RP2040 ADC does not have an on-chip reference; it uses its own power supply as a reference. On Pico W the ADC_AVDD pin (the ADC supply) is generated from the SMPS 3.3V by using an R-C filter (201Ω into 2.2µF).

1. This solution relies on the 3.3V SMPS output accuracy
2. Some PSU noise will not be filtered
3. The ADC draws current (about 150µA if the temperature sense diode is disabled, which can vary between chips); there will be an inherent offset of about 150µA*200 = ~30mV. There is a small difference in current draw when the ADC is sampling (about +20µA), so that offset will also vary with sampling as well as operating temperature.

Changing the resistance between the ADC_VREF and 3.3V pin can reduce the offset at the expense of more noise, which is helpful if the use case can support averaging over multiple samples.

Driving the SMPS mode pin (WL_GPIO1) high forces the power supply into PWM mode. This can greatly reduce the inherent ripple of the SMPS at light load, and therefore reduces the ripple on the ADC supply. This does reduce the power efficiency of the Pico W at light load, so at the end of an ADC conversion PFM mode can be re-enabled by driving WL_GPIO1 low once more. See Section 3.4.

The ADC offset can be reduced by tying a second channel of the ADC to ground, and using this zero measurement as an approximation to the offset.

For much improved ADC performance, an external 3.0V shunt reference, such as LM4040, can be connected from the ADC_VREF pin to ground. Note that if doing this the ADC range is limited to 0V - 3.0V signals (rather than 0V - 3.3V), and the shunt reference will draw continuous current through the 200Ω filter resistor (3.3V - 3.0V)/200 = ~1.5mA.

Note that the 1Ω resistor on Pico W (R9) is designed to help with shunt references that would otherwise become unstable when directly connected to 2.2µF. It also ensures there is filtering even in the case that 3.3V and ADC_VREF are shorted together (which users who are tolerant to noise and want to reduce the inherent offset may wish to do).

R7 is a physically large 1608 metric (0603) package resistor, so can be removed easily if a user wants to isolate ADC_VREF and make their own changes to the ADC voltage, for example powering it from an entirely separate voltage (e.g. 2.5V). Note that the ADC on RP2040 has only been qualified at 3.0/3.3V, but should work down to about 2V.

3.4. Powerchain

Pico W has been designed with a simple yet flexible power supply architecture and can easily be powered from other sources such as batteries or external supplies. Integrating the Pico W with external charging circuits is also straightforward. Figure 6 shows the power supply circuitry.

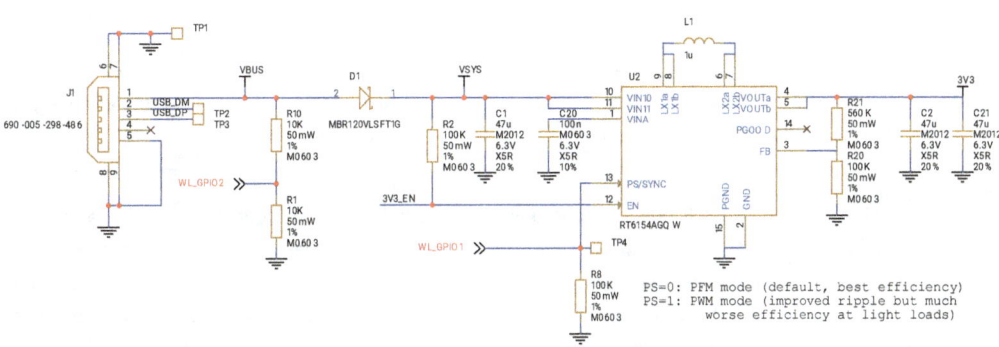

Figure 6. The powerchain of the Pico W Rev3 board.

VBUS is the 5V input from the micro-USB port, which is fed through a Schottky diode to generate VSYS. The VBUS to VSYS diode (D1) adds flexibility by allowing power ORing of different supplies into VSYS.

VSYS is the main system 'input voltage' and feeds the RT6154 buck-boost SMPS, which generates a fixed 3.3V output for the RP2040 device and its I/O (and can be used to power external circuitry). VSYS divided by 3 (by R5, R6 in the Pico W schematic) and can be monitored on ADC channel 3 when a wireless transmission isn't in progress. This can be used for example as a crude battery voltage monitor.

The buck-boost SMPS, as its name implies, can seamlessly switch from buck to boost mode, and therefore can maintain an output voltage of 3.3V from a wide range of input voltages, ~1.8V to 5.5V, which allows a lot of flexibility in the choice of power source.

WL_GPIO2 monitors the existence of VBUS, while R10 and R1 act to pull VBUS down to make sure it is 0V if VBUS is not present.

WL_GPIO1 controls the RT6154 PS (power save) pin. When PS is low (the default on Pico W) the regulator is in pulse frequency modulation (PFM) mode, which, at light loads, saves considerable power by only turning on the switching MOSFETs occasionally to keep the output capacitor topped up. Setting PS high forces the regulator into pulse width modulation (PWM) mode. PWM mode forces the SMPS to switch continuously, which reduces the output ripple considerably at light loads (which can be good for some use cases) but at the expense of much worse efficiency. Note that under heavy load the SMPS will be in PWM mode irrespective of the PS pin state.

The SMPS EN pin is pulled up to VSYS by a 100kΩ resistor and made available on Pico W pin 37. Shorting this pin to ground will disable the SMPS and put it into a low power state.

> **NOTE**
>
> The RP2040 has an on-chip linear regulator (LDO) that powers the digital core at 1.1V (nominal) from the 3.3V supply, which is not shown in Figure 6.

3.5. Powering Raspberry Pi Pico W

The simplest way to power Pico W is to plug in the micro-USB, which will power VSYS (and therefore the system) from the 5V USB VBUS voltage, via D1 (so VSYS becomes VBUS minus the Schottky diode drop).

If the USB port is the **only** power source, VSYS and VBUS can be safely shorted together to eliminate the Schottky diode drop (which improves efficiency and reduces ripple on VSYS).

If the USB port is **not** going to be used, it is safe to power Pico W by connecting VSYS to your preferred power source (in the range ~1.8V to 5.5V).

> **IMPORTANT**
>
> If you are using Pico W in USB host mode (e.g. using one of the TinyUSB host examples) then you must power Pico W by providing 5V to the VBUS pin.

The simplest way to safely add a second power source to Pico W is to feed it into VSYS via another Schottky diode (see Figure 7). This will 'OR' the two voltages, allowing the higher of either the external voltage or VBUS to power VSYS, with the diodes preventing either supply from back-powering the other. For example a single Lithium-Ion cell* (cell voltage ~3.0V to 4.2V) will work well, as will three AA series cells (~3.0V to ~4.8V) and any other fixed supply in the range ~2.3V to 5.5V. The downside of this approach is that the second power supply will suffer a diode drop in the same way as VBUS does, and this may not be desirable from an efficiency perspective or if the source is already close to the lower range of input voltage allowed for the RT6154.

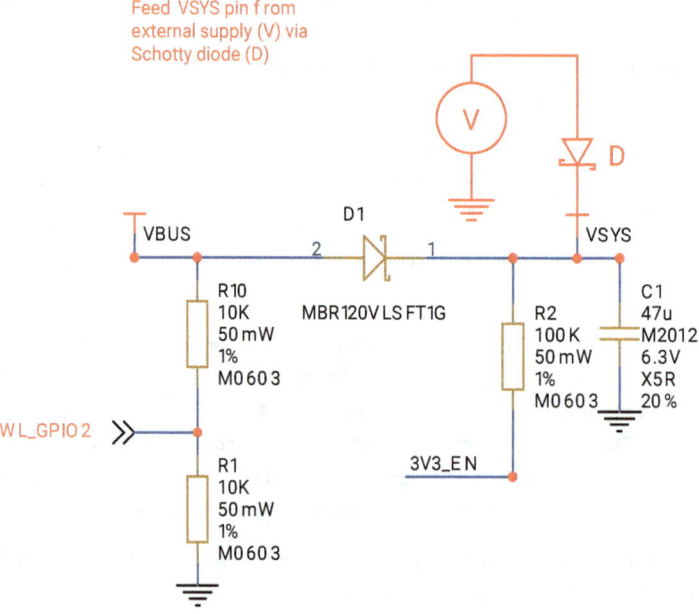

Figure 7. Pico W power ORing using diodes.

An improved way to power from a second source is using a P-channel MOSFET (P-FET) to replace the Schottky diode as shown in Figure 8. Here, the gate of the FET is controlled by VBUS, and will disconnect the secondary source when VBUS is present. The P-FET should be chosen to have low on resistance, and therefore overcomes the efficiency and voltage-drop issues with the diode-only solution.

Note that the V_t (threshold voltage) of the P-FET must be chosen to be well below the minimum external input voltage, to make sure the P-FET is turned on swiftly and with low resistance. When the input VBUS is removed, the P-FET will not start to turn on until VBUS drops below the P-FET's V_t, meanwhile the body diode of the P-FET may start to conduct (depending on whether V_t is smaller than the diode drop). For inputs that have a low minimum input voltage, or if the P-FET gate is expected to change slowly (e.g. if any capacitance is added to VBUS) a secondary Schottky diode across the P-FET (in the same direction as the body diode) is recommended. This will reduce the voltage drop across the P-FET's body diode.

An example of a suitable P-MOSFET for most situations is Diodes DMG2305UX which has a maximum V_t of 0.9V and R_{on} of 100mΩ (at 2.5V V_{gs}).

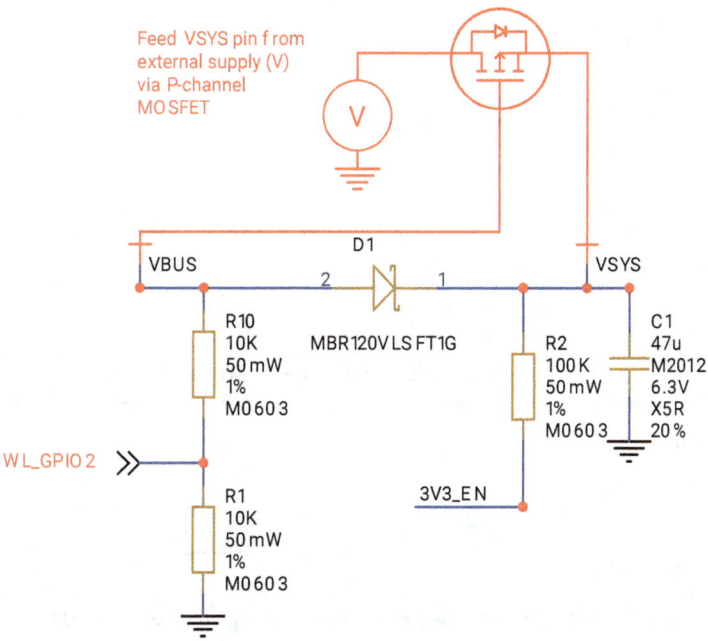

Figure 8. Pico W power ORing using P channel MOSFET.

> ⚠ **CAUTION**
>
> If using Lithium-Ion cells they must have, or be provided with, adequate protection against over-discharge, over-charge, charging outside allowed temperature range, and overcurrent. Bare, unprotected cells are dangerous and can catch fire or explode if over-discharged, over-charged or charged / discharged outside their allowed temperature and/or current range.

3.6. Using a battery charger

Pico W can also be used with a battery charger. Although this is a slightly more complex use case it is still straightforward. Figure 9 shows an example of using a 'power path' type charger (where the charger seamlessly manages swapping between powering from battery or powering from the input source and charging the battery, as needed).

Figure 9. Using Pico W with a charger.

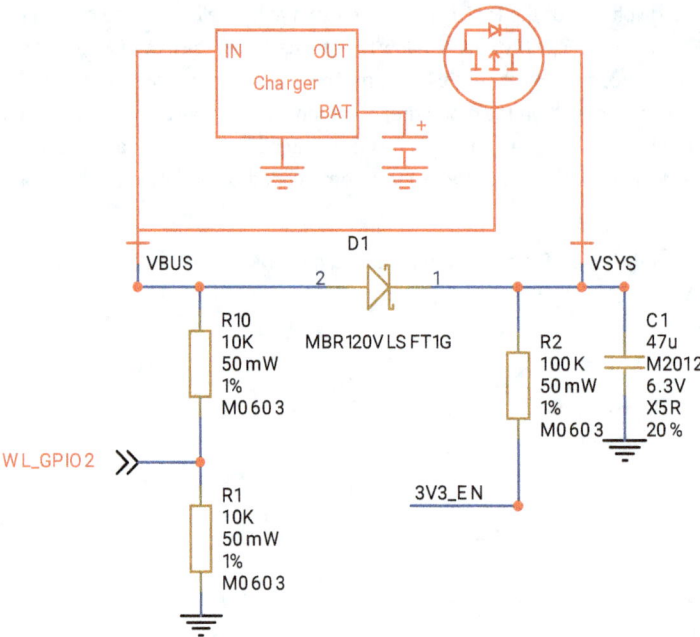

In the example we feed VBUS to the input of the charger, and we feed VSYS with the output via the previously mentioned P-FET arrangement. Depending on your use case you may also want to add a Schottky diode across the P-FET as described in the previous section.

3.7. USB

RP2040 has an integrated USB1.1 PHY and controller which can be used in both device and host mode. Pico W adds the two required 27Ω external resistors and brings this interface to a standard micro-USB port.

The USB port can be used to access the USB bootloader (BOOTSEL mode) stored in the RP2040 boot ROM. It can also be used by user code, to access an external USB device or host.

3.8. Wireless interface

Pico W contains an on-board 2.4GHz wireless interface using the Infineon CYW43439, which has the following features:

- WiFi 4 (802.11n), Single-band (2.4 GHz)
- WPA3
- SoftAP (Up to 4 clients)

The antenna is an onboard antenna licensed from ABRACON (formerly ProAnt). The wireless interface is connected via SPI to the RP2040.

Due to pin limitations, some of the wireless interface pins are shared. The CLK is shared with VSYS monitor, so only when there isn't an SPI transaction in progress can VSYS be read via the ADC. The Infineon CYW43439 DIN/DOUT and IRQ all share one pin on the RP2040. Only when an SPI transaction isn't in progress is it suitable to check for IRQs. The interface typically runs at 33MHz.

For best wireless performance, the antenna should be in free space. For instance, putting metal under or close by the antenna can reduce its performance both in terms of gain and bandwidth. Adding grounded metal to the sides of the antenna can improve the antenna's bandwidth.

There are three GPIO pins from the CYW43439 that are used for other board functions and can easily be accessed via the SDK:

WL_GPIO2

    IP VBUS sense - high if VBUS is present, else low

WL_GPIO1

    OP controls the on-board SMPS power save pin (Section 3.4)

WL_GPIO0

    OP connected to user LED

NOTE

> Full details of the Infineon CYW43439 can be found in the datasheet.

3.9. Debugging

Pico W brings the RP2040 serial wire debug (SWD) interface to a three-pin debug header. To get started using the debug port see the Debugging with SWD section in the Getting started with Raspberry Pi Pico book.

NOTE

> The RP2040 chip has internal pull-up resistors on the SWDIO and SWCLK pins, both nominally 60kΩ.

Appendix A: Availability

Raspberry Pi guarantee availability of the Raspberry Pi Pico W product until at least January 2028.

Support

For support see the Pico section of the Raspberry Pi website, and post questions on the Raspberry Pi forum.

Ordering code

Table 1. Part Number

Model	Order Code	EAN	Minimal Order Quantity	RRP
Raspberry Pi Pico W	SC0918	5056561803173	1+ pcs / Bulk	US$6.00
Raspberry Pi Pico WH	SC0919	5056561800196	1+ pcs / Bulk	US$7.00

 NOTE

RRP was correct at time of publication and excludes taxes.

Appendix B: Pico W schematic

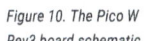

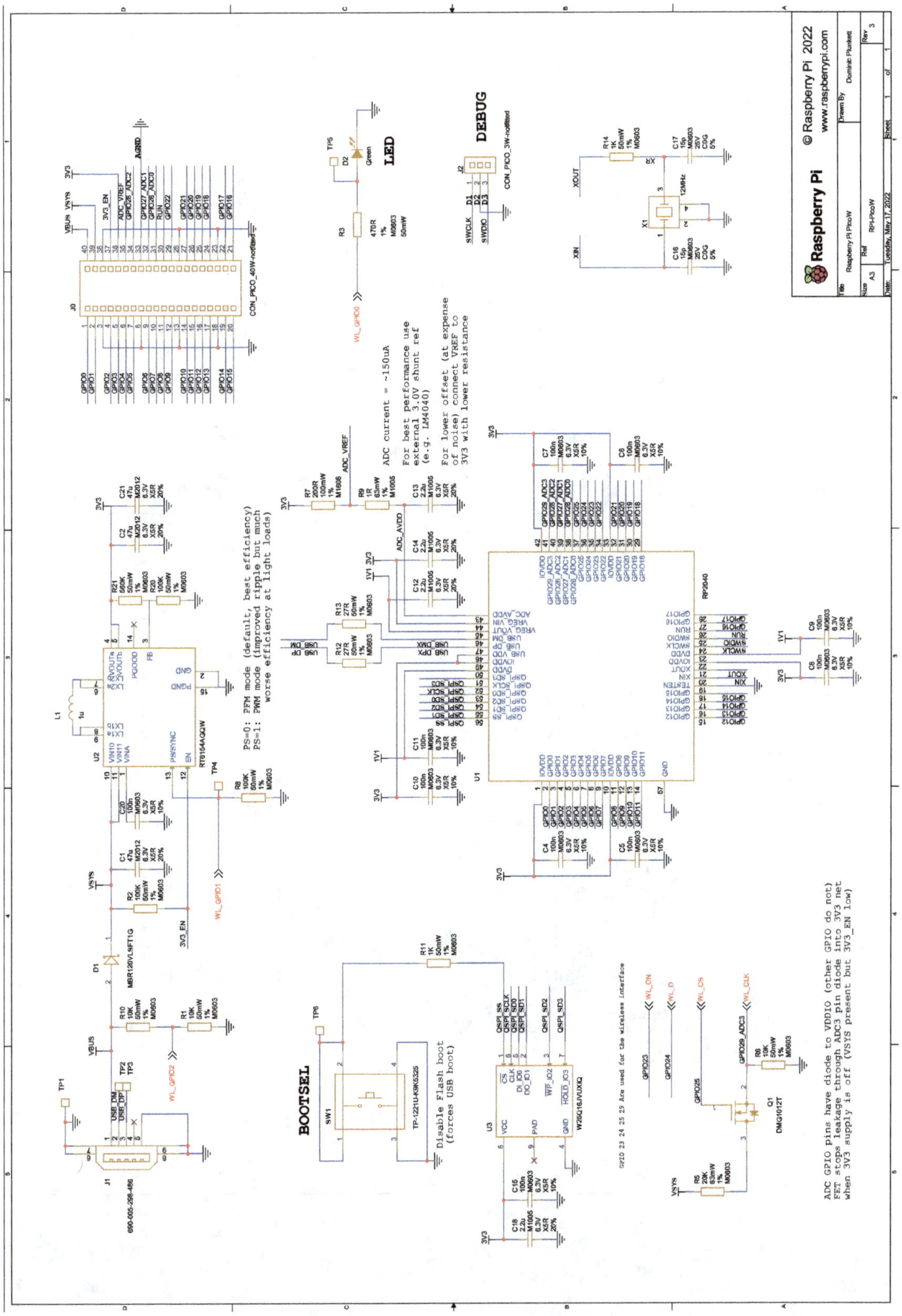

Figure 10. The Pico W Rev3 board schematic.

Raspberry Pi Pico W Datasheet

Figure 11. The Pico W Rev3 board schematic.

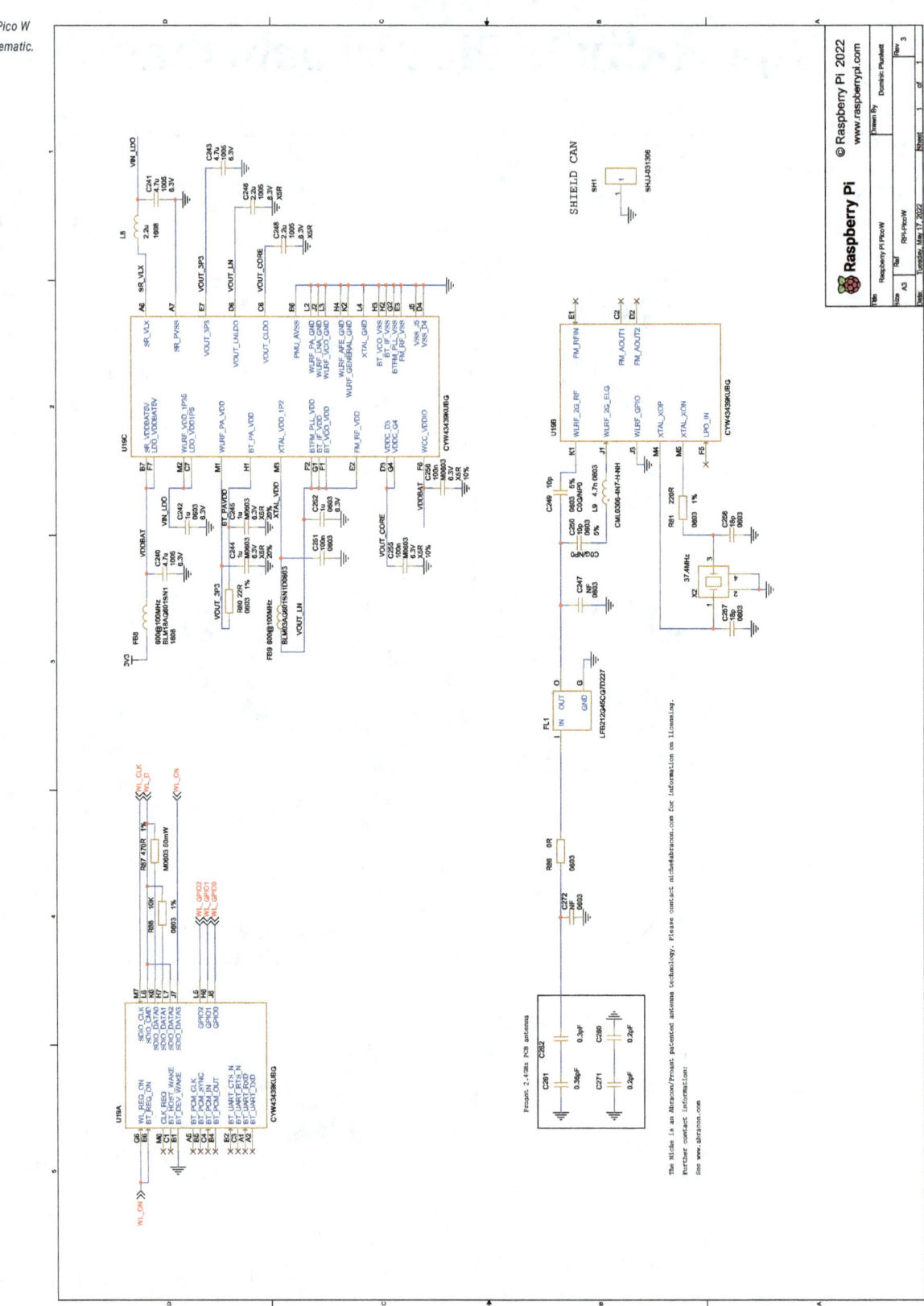

Appendix B: Pico W schematic

Appendix C: Pico W component locations

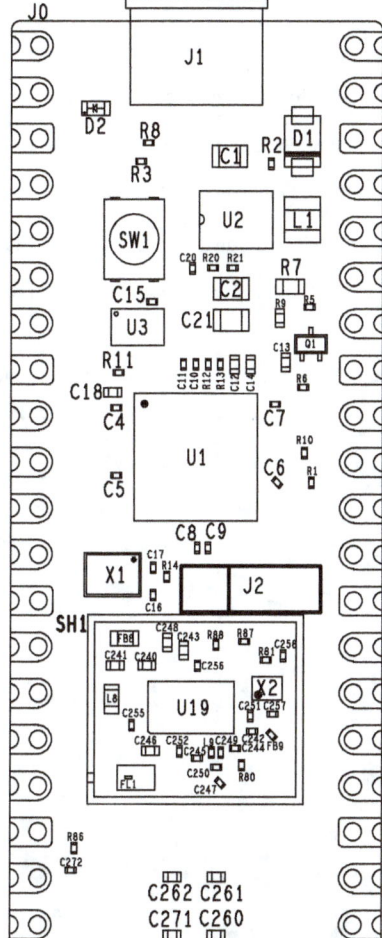

Figure 12. The Pico W Rev3 board component locations.

Appendix D: Documentation release history

Table 2. Documentation release history

Release	Date	Description
1.0	30 Jun 2022	• Initial release
Pico and Pico W databooks combined into a unified release history		
2.0	01 Dec 2022	• Minor updates and corrections • Added RP2040 availability information • Added RP2040 storage conditions and thermal characteristics • Replace SDK library documentation with links to the online version • Updated Picoprobe build and usage instructions

The latest release can be found at https://datasheets.raspberrypi.com/picow/pico-w-datasheet.pdf.

Raspberry Pi is a trademark of Raspberry Pi Ltd

www.ingramcontent.com/pod-product-compliance
Lightning Source LLC
Chambersburg PA
CBHW081052170526
45158CB00007B/1951